CLOWNFISHES

(Anemonefishes of the Genera *Amphiprion* and *Premnas*)

Richard F. Stratton

THE APPEAL OF CLOWNFISHES

If ever there was a fish that could compete simultaneously in the categories of "most beautiful," "most interesting," and "most exotic," it is the clownfish—although just which species of clownfish would be open to debate. There are twenty-six known species of clownfish, and at least ten of them would contest for the "most beautiful" title.

Part of the beauty lies in the contrasting colors. So often I have observed people seeing living clownfish for the first time commenting upon how the white stripes looked as though they had been painted upon the beautiful fish. And part of the esthetic appeal results from the graceful movements of the clownfishes. Few fishes have such elegance of manner.

Dedicated to: Veronica

© T.F.H. Publications, Inc.

Distributed in the UNITED STATES to the Pet Trade by T.F.H. Publications, Inc., 1 TFH Plaza, Neptune City, NJ 07753; on the Internet at www.tfh.com; in CANADA by Rolf C. Hagen Inc., 3225 Sartelon St., Montreal, Quebec H4R 1E8; Pet Trade by H & L Pet Supplies Inc., 27 Kingston Crescent, Kitchener, Ontario N2B 2T6; in ENGLAND by T.F.H. Publications, PO Box 74, Havant PO9 5TT; in AUSTRALIA AND THE SOUTH PACIFIC by T.F.H. (Australia), Pty. Ltd., Box 149, Brookvale 2100 N.S.W., Australia; in NEW ZEALAND by Brooklands Aquarium Ltd., 5 McGiven Drive, New Plymouth, RD1 New Zealand; in SOUTH AFRICA by Rolf C. Hagen S.A. (PTY.) LTD., P.O. Box 201199, Durban North 4016, South Africa; in JAPAN by T.F.H. Publications. Published by T.F.H. Publications, Inc.

MANUFACTURED IN THE
UNITED STATES OF AMERICA
BY T.F.H. PUBLICATIONS, INC.

The claim for the "most interesting" title springs from several aspects of the animal, but a major component of the claim is the remarkable relationship that the clownfish species have evolved with the giant species of sea anemones in the Indo-Pacific regions. These large anemones have nematocysts, or stinging cells, on their tentacles, yet the clownfishes live with impunity among those deadly tentacles and are thus protected from would-be predators. This association was reported over a hundred years ago, but humankind has never ceased to wonder at it. We will talk more about the nature of this association later.

Clownfishes are also particularly interesting because of their reproductive patterns. Again, we will discuss the natural history of these animals in more detail later. For now it is sufficient to say that clownfishes live as mated pairs, sometimes with a few juveniles in the same anemone, and lay eggs that both parents guard and protect until they hatch. Once the eggs hatch, the young are dispersed by the parental fish, aided by the currents, to the planktonic rafts, in which the young will spend at least two weeks. An important point here is that the fish live as a truly mated pair. Many hobbyists will think of cichlids in this respect, but the truth is that not many species live as pairs. Although the cichlids are masters of parental care, most species pair up only to raise the young to a certain point, and then the pair part company. (That little-known fact is the reason so many

CONTENTS

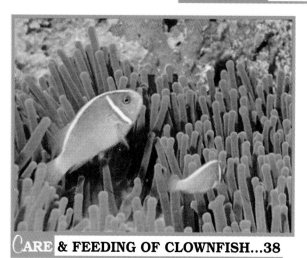

cichlid hobbyists have compatibility problems with their "mated pairs" after the young are removed.)

As for the "most exotic," it may very well be that the various different species are too beautiful and ordinary looking (in the sense that they don't have a bizarre shape) to win that title, and yet they are certainly quite exotic. For one thing, their distribution is limited. They are found in the Indian Ocean and in the Pacific as far east as Tahiti, but that is pretty much it. Their restricted range, great beauty, and fascinating life history certainly qualify them for the term exotic, even if they don't win the category based on bizarre shape.

The main point is that clownfish are something special, and it is not by accident that they have been the object of the attention of marine aquarists for many years. Nor is it by mere coincidence that they were among the first marine fish species to be bred and raised successfully in captivity. The fact is that next to the seahorse, the clownfish is the most emblematic of the sea of all fish species. (The seahorse, on the other hand, is an example of a definitely exotic, not to mention charming, fish that would not quite be called "most beautiful." Not even the red ones!)

So compelling has the charm of clownfishes been that the naming of the group has been lost in antiquity. But it doesn't take a genius to see a similarity between clown costumes and the coloration of certain species, such as the so-called common clownfishes *Amphiprion percula* and

Amphiprion percula.
Photo by Dr. Herbert R. Axelrod.

Amphiprion ocellaris and the maroon clownfish, *Premnas biaculeatus.* Some have speculated that the behavior of the clownfishes is also clownlike. Certainly they are particularly bold because they have the protection of the anemone, and it is amusing to see them rub against the anemone, much like a hungry cat rubbing against its owner's legs. Nevertheless, it seems clear that the name clownfish originated with the similarity to a clown's costume.

It is not surprising that the clownfishes are also known as "anemonefishes." They are, perhaps, more often referred to in that way among ichthyologists and other scientists, such as ethologists and sociobiologists. This term is more inclusive and fitting in some ways, since some clownfish species, such as *Amphiprion ephippium* and *Amphiprion akallopisos,* have no vertical

Amphiprion ocellaris.
Photo by Richard T. Bell courtesy of The Fish Store, Bellevue, Wa.

Amphiprion akallopisos from the Seychelles.
Photo by Dr. John Randall.

group of fishes, while anemonefishes and amphiprions are also common appellations for the group. Scientifically, they are known as the subfamily Amphiprioninae, but even scientists don't often refer to them as amphiprionids, partly because "clownfish" is firmly established as a popular name.

To understand clownfishes better, it is necessary to know their taxonomic relationship to other fish species. The clownfishes are members of the damselfish family

stripes at all, and others have only one or two. But the clownfish name has been around long enough that it is safe to say that it is here to stay. And the name "anemonefishes" has the disadvantage of the presumptive inclusion of *Dacyllus trimatulatus*, since the juveniles of this species also utilize anemones for protection, although they are not nearly as evolved and specialized in this behavior as are the clownfish species.

Fishes in the clownfish group are sometimes referred to collectively as the "amphiprions," which is a case of identifying the group from the generic name *Amphiprion*. Although this term is utilized by both hobbyists and scientists, it has the disadvantage of not including the genus *Premnas*, of which *Premnas biaculeatus* is a member and very much a clownfish. So the situation is that clownfish is the most-used name for this amazing

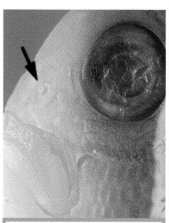

A preserved *Amphiprion* showing the single nostril. Photo by Dr. Melanie L.J.Stiassny.

Pomacentridae, and the fact is that a non-clownfish damsel, *Dacyllus trimaculatus*, makes use of the protection of anemones, but it is only the juveniles which do so. The family Pomacentridae has been grouped by recent taxonomic research into the superfamily Labroidea, which includes wrasses, cichlids, parrotfishes, and surfperches. The basis for the grouping, among others, is that the fishes of each of these groups have teeth in the throat

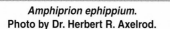

Amphiprion ephippium.
Photo by Dr. Herbert R. Axelrod.

Amphiprion nigripes, one of the more spectacular of the clownfishes. The darkness of the ventral and anal fins contrasts with the lightness of the body.
Photo by Dr. Herbert R. Axelrod

(pharyngeal teeth) as well as teeth in the jaws. Other groups of fishes have one or the other, but not both.

That some scientists group the clownfishes into a sub-family, Amphiprioninae makes sense. Clownfishes are definitely damsels, but they are very specialized ones. Damsels are known as some of the most territorial and aggressive of all fishes in the sea. It is surprising that these colorful little fishes are such tyrants, but the behavior is tied to the fact that many species feed off algae, and some even "farm," cultivating certain algal species by eliminating competing algal species, and it is in their interest to drive off animals that would eat "their" patch of algae.

Clownfishes generally are not as aggressive as many other damsels. However, most coral reef species have some degree of aggression, and we don't get off scot-free with clownfishes either. Clownfishes' aggression can be easily managed, though. It is somewhat ironic that damsels have such a meek-sounding name, as not only are they ounce-per-ounce about as tough as a fish gets, but they are quite durable, too. Because of the hardiness of damsels and the fact that they are usually less expensive than many other species, aquarists often use them to "break in" a tank. That is, they are used to "feed" the beneficial bacteria in the biological filters with their metabolic byproducts. And they are expected to survive the violent fluctuations in water chemistry that occur before the biological filter begins to function. Once the biological filter has begun to perform well, it is often amusing that the hobbyist is then unable to put in any "good" or "exotic" fish because the damsels have taken over the entire tank as their territory and will not allow any "intruders."

Dascyllus trimaculatus.
Photo by Andre Roth.

Amphiprion bicinctus.
Photo by Wallace Heaton.

One of the advantages of clownfishes is that they can be kept with other damsels, particularly if they are kept with an anemone. Coral reef fishes know what an anemone is, and they stay away from the stinging tentacles. The clownfish will be safe and secure in the anemone, even though they make occasional excursions away from it, and the other damsels will take up residence in the coral rubble of the tank. There are many species of blue damsels which go particularly well with clownfishes, but the fact is that the clownfish with an anemone can be kept with almost any fish species, for the anemone will provide safety.

While the clownfishes are not quite as aggressive as most other damsels, it should also be noted that they are not quite as "bulletproof" as many of the damsel species that are utilized as a first fish by many marine hobbyists. And they don't ship as well as many of those other damsels. The good news is that some clownfishes are available commercially now as "tank raised." That means that the fish don't have to be captured in the wild and shipped, for commercial breeders and hobbyists routinely breed many of these species now. I say "routinely," but it should be noted that they are still quite challenging to breed and raise. The point is, though,

that only a couple of decades ago it was thought to be impossible to raise nearly any coral reef species. That was because of the pelagic larval form that nearly every species has. The fact that clownfishes can now be bred and raised commercially or as a hobby is a coming of age of sorts for the marine hobby. And it is a further appeal of clownfishes.

CLOWNFISHES AND ANEMONES

Since many aquarists start right out with saltwater fish these days (something that would have been thought to be very indvisable, if not almost impossible, just a few decades ago), I am going to explain everything. If I leave

A magnificent display of
clownfishes in their wild habitat.
Photo by Cathy Church.

something out, it is from a personal lapse. Anemones are sea animals that could be called upside-down jellyfish. Jellyfish hang in the water with the stinging tentacles hanging down. Anemones have a muscular "foot," much like a snail, which they use to attach themselves to the substrate and move about. Their stinging tentacles form a crown around the oral opening to the stomach. And they point upward.

Although the stinging tentacles are capable of killing many fish, they are not dangerous to humans. I have touched nearly every species, and they merely feel "sticky" from the firing of the nematocysts. The tough skin of the human hand is normally ample protection.

Individuals can be allergic to the stinging cells and may develop a mild rash, but this is rare. Even among those with allergies, the rash is more likely to occur in more sensitive areas of the skin on the forearm or the back of the hand.

While the clownfishes were named after clowns, the sea anemone was named after the anemone flower. It is ironic that the sea anemone may now be better known among the general public than the flower, but I speak as one who lives near the ocean.

The sea anemones that the clownfishes inhabit are generally giant anemones found in the shallower areas of the coral jungles. Clownfishes vary in how particular they are about the

anemone they will accept. When discussing individual species, I'll make note of the anemones they accept, but I should mention here that some species are so unparticular that they will even pester corals to death rubbing against them. But they only do this if there is no anemone in the tank. The anemones take a long time to grow to the size that they reach in the wild (about three feet across for the largest ones), so it is possible to get one of the smaller species and keep it for many years before it gets too big for a small tank. It should be mentioned, however, that sea anemones are much more difficult to keep than clownfishes. They have very special requirements.

Amphiprion clarkii **spends a lot of time in the protection of an anemone when it is even slightly disturbed.**

The giant clam of the South Seas, *Tridacna gigas*. Photographed in the Maldive Islands. Photo by Dr. Herbert R. Axelrod.

For many years, the tropical anemones that the clownfishes accepted could not be kept by hobbyists. There were two reasons for this. One was that sea anemones of this type have an obligatory relationship with zooxanthellae that live in their tissues. These zooxanthellae are actually flagellates, but they can be thought of as specialized microscopic organisms that are able to perform photosynthesis. In fact, they are often referred to as symbiotic algae, even though they technically are not algae. They are highly evolved flagellates with chloroplasts, and they lose the flagella once they are ensconced in the tissues of the host.

The problem with keeping such anemones is that they have a need for very strong lighting and for actinic lighting. Actinic lighting is lighting that can cause a physical or chemical change in an organism, such as photosynthesis—and sunburn! Today such lighting is available, and it is conveniently packaged for marine aquaria that are to be used for keeping photosynthetic organisms, such as anemones and corals. (There are also photosynthetic clams. In fact, the giant clam of the South Seas, *Tridacna gigas*, is such a species.)

In addition to the special lighting, anemones have a need for super clean water. Such water can be maintained by the use of biological filtration and a protein skimmer. The biological filtration can consist of trickle filtration or fluidized bed filtration or both. The fact is that these filters complement each other quite well.

Although an anemone with clownfish is not only one of the most beautiful displays possible in an aquarium and one of the most interesting, it should be mentioned that clownfishes can be kept without an anemone. In fact, many clownfish species have even spawned without an anemone present in the tank. Most hobbyists who particularly like clownfishes are going to want to keep them with an anemone, but it is worth making the point that the many different species are also popular in fish-only marine community aquaria.

For many years, ichthyologists were at a loss to understand how the clownfish species were able to survive in the stinging tentacles of the sea anemone. What made this especially mysterious is that the clownfishes inhabited the largest and most deadly of the anemones. The speculation was that the fishes had a very specialized body slime that made them immune to the stings of the coelenterates. Research indicated that, true enough, the clownfishes did have a specialized body slime, but it didn't make them immune to the stinging nematocysts. Instead, it made them "invisible" to the anemone. What happens is that clownfish "flirt" with an anemone, enduring a few stings, but each time they touch the anemone, the specialized body slime takes on some molecules of the body of the anemone. This causes the anemone to fail to sense the fish as a foreign presence, so its stinging cells, the nematocysts, do not fire. And a clownfish must constantly rub against the anemone in order to keep its body slime sufficiently covered with anemone molecules to provide the "scent."

The relationship of the clownfish and the anemone is usually referred to as symbiotic, meaning that each organism benefits. The benefit for the clownfish is obvious. It receives protection from predators. But what is the benefit for the anemone? The truth is that there are many anemones that live in their natural habitat without clownfishes, so the relationship is not an obligatory one for them. But there are no clownfishes seen in the wild without anemone hosts, so it is most definitely an essential relationship for the fishes.

Is there any benefit for the anemone? Cousteau filmed clownfish feeding the anemone in the wild. The way that worked was that divers gave food to the clownfish, and the clownfish carried the food back to the anemone and stuffed it in among its tentacles, from which it was later transferred to the mouth. In some cases, the fish even shoved the food into the mouth.

Amphiprion clarkii.
Photo by Klaus Paysan.

Amphiprion polymnus looks like a real clown. this was the first *Amphiprion* to be described (Linnaeus 1758), and may very well be the most attractive, but it is unknown in the hobby.
Photo by Dr. Herbert R. Axelrod

Perhaps the greatest authority on clownfishes, Dr. Gerald R. Allen, disputes that this is typical clownfish behavior. Also an expert on damselfishes, he maintains that the storing of food is a typical pomacentrid trait. So although it makes a pretty story, clownfish don't really feed the anemone—at least not on purpose. The fact is that there is no need. The currents wash the anemone with plankton, and the symbiotic zooxanthellae provide the anemone with food, too.

Again, we are back at square one. Does the anemone benefit from the association? Clownfish are constantly observed cleaning the anemone. This is important to sea anemones, but currents do that job, too, so it is not obligatory for the clownfish to be present. Still, every little bit helps. The primary advantage to the anemone of having "tenants" probably derives from the fact that clownfish defend the anemone against those predators that would prey upon its tentacles. These predators range from fishes to crustaceans to mollusks.

Beautiful almost beyond compare, the appeal of clownfish is obvious. The anemone is so exotic it is almost alien, and appreciating its beauty may be an acquired taste. In the case of anemones and clownfish, though, the effect of the combination is startling. Such a display is breathtaking and full of fascination. Now that such displays can be maintained, it is of little wonder that the clownfishes are more popular than ever. That's okay. They deserve to be. Now let's take a look at each of the known clownfish species.

Amphiprion melanopus.
Photo by Klaus Paysan.

CLOWNFISH SPECIES

Although there are presently twenty-three known species of clownfish, it is difficult to say which is the most beautiful. There is even disagreement among hobbyists about which is the most easily kept or easily spawned. The most popular clownfishes have been the common clownfish species. But some hobbyists prefer the appearance of the tomato clownfish, *Amphiprion frenatus*, even though it must be confessed that *Amphiprion rubrocinctus* looks very much like it. In any case, I'll give you some information about each species, including its popularity, its distribution in the wild, and the anemones that it is known to accept.

SCIENTIFIC NAMES

There are a number of reasons that we are going to get used to some of the scientific names for the different clownfish species. Because of that, I would like to give some basic guidelines for understanding and using them. That doesn't mean that we won't make use of the popular names, too. After all, even ichthyologists make use of them; when they describe a species, they often suggest a popular name, too, but of course its use is not mandatory.

One of the problems with popular names is that the names are not always applied in the same way by everyone around the world—or even across a country, for that matter. An example is the name "common clownfish."

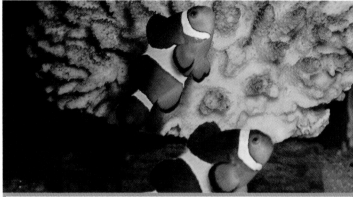
Amphiprion ocellaris.
Photo by Klaus Paysan.

Amphiprion perideraion photographed in Marau Island, British Solomons.
Photo by Dr. Herbert R. Axelrod.

Premnas biaculeatus from the Philippine Islands.
Photo by Gene Wolfsheimer.

That name was applied to *Amphiprion ocellaris* in the United States, but in Australia it was applied to *Amphiprion percula*.

Another point worth mentioning is that when a species gets its common name the name of choice is mere happenstance, but when the scientific name is applied, a qualified researcher utilizes preserved specimens. He (or she) counts scales, counts fin rays, makes measurements, and dissects the fish to find traits that qualify it as a separate species. An exhaustive search is then made of the scientific literature to compare all measurements, counts, and physical traits with the specimen in question. In spite of all of this, there is occasionally some confusion that arises with scientific names, but not nearly equal to that of popular names. And the fish is known around the world by the scientific name. Since there would be dissension and hurt feelings if a living language were utilized, Latin is the preferred language. In fact, the term "Latin name" is often used synonymously with the scientific name. It should be mentioned, though, that some Greek terms are thrown in.

Although the scientific names are written the same around the world, they are not pronounced the same. As many ichthyologists have observed with amusement, the guy who says the name with the most confidence usually prevails. The truth is that there were once strict rules for pronunciation, but the International Nomenclature Association gave up to some degree on trying to enforce any of them when it eschewed diacritical marks in any of the names. Generally speaking, however, "ae" is a diphthong that is pronounced "ee." For that reason, the

Amphiprion ocellaris.

Amphiprion clarkii.
Photo by Karl Knaack.

family names, such as Pomacentridae, all end in an "ee" sound.

Names that should never be mispronounced (but often are) are the patronyms. These are cases in which the fish has been named after a particular person. For example, if a clownfish were to be named after a man named Allard, the name is given as *allardi*, which is Allard with an "i" added. The name is simply pronounced as the man's name would normally be pronounced, with "i" (pronounced "eye") at the end. In earlier days, it was common practice to place two "i"s at the end of the name. An example is *Amphiprion clarkii*. Obviously, the fish was named after someone named Clark, so the name is simply

pronounced "clark," and the first "i" is pronounced ee, but the second one is pronounced normally. So it is pronounced "clark-ee-EYE."

Difficulties arise when species are named after names in a different language, such as *agassizi* and *ramirezi*. Obviously, some mistakes will be made if we aren't familiar with how the name is pronounced, but at least it will minimize confusion if we remember to pronounce the "i" ending a patronym as "eye."

A common error is to capitalize the species name. *Amphiprion Clarkii* is incorrect. It should be *Amphiprion clarkii*. Also, scientific names are underlined or placed in italics to denote that they are a foreign language. Finally, it

should be mentioned that the scientific name is a proper noun. Hence, we would say, "*Amphiprion ocellaris* is a pretty fish." "The *Amphiprion ocellaris* is a pretty fish" would be incorrect, and this is a very common error among hobbyists.

Just to place everything in perspective, all living things are classified in a kingdom, a phylum, a class, an order, a family, a genus, and a species. Thus, humans are in the animal kingdom, the chordate phylum, the class of mammals, the primate order, and our own family, Hominidae. (Hey, we made up the classification, didn't we? So we get our own family. If we were a lesser species studied by something else more objective, we might be grouped with the apes.) Our genus is *Homo*, and our species is *sapiens* (meaning "wise," another advantage of doing the classifying!). From now on, notice how often newspapers and other publications fail to place *Homo sapiens* in italics or how they fail to capitalize the generic name (or capitalize both names). They usually manage to use it improperly in some way.

Biology students learn little sentences to help them remember the different levels of classification, with the first letter standing for each level. A common one is "Kings play chess on fancy gilded sets." This helps us remember Kingdom, Phylum, Class, Order, Family, Genus, and Species. It should be mentioned that different levels can be inserted, such as superfamily or subfamily, and this can be done at any level by scientists doing more detailed work.

Now that we have cut our teeth on scientific names, it is time to actually look at some of our species of clownfish—at long last! We will go in alphabetical order by scientific name.

Amphiprion akallopisos Bleeker 1853

I waited until now to tell you that in many cases in which a particular species is mentioned, the name of the original describer of the species is placed after the name, along with the date that the description was published. The popular name for this fish is skunk clownfish because of the white stripe that runs down its back. The species is a popular and hardy aquarium fish in

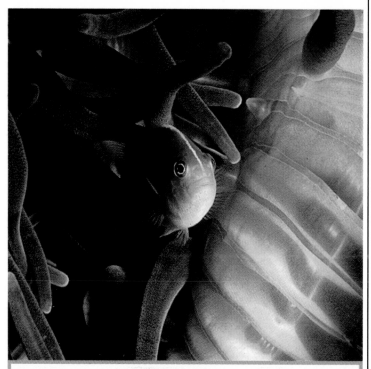

Amphiprion akallopisos.
Photo by MP&C Piednoir.

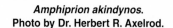

Amphiprion akindynos.
Photo by Dr. Herbert R. Axelrod.

spite of the fact that its coloration doesn't measure up to that of many of the others. It is found throughout the tropical areas of the Indian Ocean. This species accepts only the anemone *Heteractis magnifica.*

Amphiprion akindynos Allen 1972

There is no popular name established in the hobby in the United States for this species, as we have so seldom seen it. It is primarily found on the Barrier Reef of Australia, and collecting there is strongly controlled. Dr. Gerald R. Allen (who described the species) suggests the popular name of Barrier Reef clownfish. Hobbyists fortunate enough to get this species may simply utilize the scientific name. *A. akindynos* will accept several anemones, including *Entacmaea quadricolor, Heteractis crispa, Heteractis magnifica,* and *Stichodactyla haddoni.*

Amphiprion allardi Klausewitz 1970

Although this species was known to science from the last century, it was incorrectly confused with other clownfish species. Again, this is a species that has not been prevalent in the American hobby, as it is found along the tropical east coast of Africa, from Kenya to Durban. For that reason, there is no generally recognized American common name for the species, although the species with a patronym for a name can always be called by that as the popular name. For

Amphiprion allardi.
Photo by Dr. Herbert R. Axelrod.

Heteractis magnifica, a favorite anemone of many species of clownfishes.
Photo by U. Erich Friese.

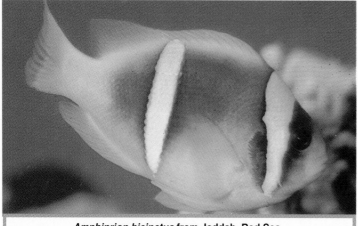

**Amphiprion bicinctus from Jeddah, Red Sea.
Photo by Dr. Gerald R. Allen.**

**Amphiprion chrysogaster.
Photo by Dr. Herbert R. Axelrod.**

**Amphiprion chrysopterus from Marau Island, British Solomons.
Photo by Dr. Herbert R. Axelrod.**

example, *A. allardi* easily becomes "Allard's clownfish or anemonefish." The species will accept the anemones *Entacmaea quadricolor*, *Heteractis aurora*, and *Stichodactyla mertensii*.

Amphiprion bicinctus Ruppel 1828

This species is endemic to the Red Sea, meaning that it is found only there. Nevertheless, it has been imported as an aquarium fish and has been fairly hardy in captivity. It is usually called the "two-banded clownfish," and it accepts the anemones *Entacmaea quadricolor*, *Heteractis aurora*, *Heteractis crispa*, and *Stichodactyla gigantea*.

Amphiprion chagosensis Allen 1972

Since this clownfish is found only on two islands or atolls (namely Diego Garcia Atoll and the Chagos Archipelago in the Indian Ocean), it is very unlikely that it has been kept as an aquarium specimen in the United States. In any event, it is quite similar in appearance to many others. This is a little-known species with little information regarding its natural history and ecology, including the anemones that it accepts. At this stage of the game, it is primarily a "museum specimen" in that it is unstudied in the wild and very likely unknown in the aquarium hobby.

Amphiprion chrysogaster Cuvier 1830

Again, this is a species with a very limited distribution, and it is probably unknown in the hobby. It is known only from the island of Mauritius in the Indian Ocean. It is symbiotic with *Heteractis aurora*, *Stichodactylus haddoni*, and *Stichodactylus mertensii*. The only common name that has been suggested is Mauritian clownfish, from its only known locality. Again, this is something of a museum species, known only from a few preserved specimens and not much studied in the wild. However, its very rarity will make it that much more exotic and desirable to many hobbyists—even if they may have to make special arrangements to collect the fish!

Amphiprion chrysopterus Cuvier 1830

This clownfish has a wide distribution, from New Guinea to the Coral Sea to the Fiji Islands. With such a wide distribution, the fish has been often seen in the aquarium hobby, although it is often misidentified. According to Allen, the specimens from Melanesia have black pelvic and anal fins, while these fins are yellow to orange throughout the remainder of the species's distribution. The common name orange-fin clownfish has been suggested, but most hobbyists and dealers who are able to identify this fish simply use the scientific name. The species accepts several different anemones, including *Heteractis aurora*, *Heteractis crispa*, and *Strichodactyla mertensii*. It has been relatively hardy in the aquarium, much like *Amphiprion clarkii*, with which it is often confused.

Amphiprion chrysopterus in their natural habitat.
Photo by Dr. Gerald R. Allen.

Amphiprion chrysopterus in its anemone, **Heteractis magnifica.** The blue cast is a result of shooting the picture in shallow water without a strobe light.
Photo by Dr. Gerald R. Allen.

Amphiprion clarkii.
Photo by Dr. Herbert R. Axelrod.

Amphiprion clarkii likes to take shelter in its anemone.

Amphiprion clarkii (Bennett 1830)

The name of the describer is placed in parentheses in the case of this species, and that is because Bennett described it under a different generic name, *Anthias;* he did that in 1830. (When the species is valid but it is later determined that it belongs in a genus other than the one under which it was first described, the name of the describer is included in parentheses; when the date of the original description in such cases is known, it too is enclosed within the parentheses.)

In any case, this is a very widespread species, and its coloration varies from orange to black. In fact, I saw black examples of this species while diving off Moorea, which is an island near Tahiti. The "official" popular name for this species is Clark's clownfish, but it is usually simply referred to as "clarkii" by hobbyists and ichthyologists. The species is hardy in the aquarium, and because of its widespread distribution it is usually available. It accepts many different anemones, including just about any that are from the South Pacific, and it has been reported to even accept anemones from the Caribbean. In the wild, it is associated with *Cryptodendrum adhesivum, Entacmaea quadricolor, Heteractis aurora, Heteractis crispa, Stichodactyla gigantea, Stichodactyla haddoni*, and *Stichodactyla mertensii.*

Amphiprion ephippium (Bloch 1790)

Nicknamed the saddlebacked clownfish, this

species is rarely seen in the hobby, as its distribution is limited to the Java Sea off Jakarta. Juveniles have white stripes that disappear as the individuals mature and develop the black saddle that can be quite variable among individuals. The species is symbiotic with *Heteractis crispa*.

Amphiprion frenatus Brevoort 1856

Popularly called the tomato clownfish, this species is very popular among hobbyists, and it is in the running for being acclaimed one of the most beautiful of the clownfishes. Youngsters have several stripes that later disappear. The species is found north to Japan, as well as Malaysia, Okinawa, Thailand, China, and the Philippines. It is symbiotic with *Entacmaea quadricolor*; however, it has been bred in captivity many times without the anemone host present.

Amphiprion fuscocaudatus Allen 1972

This species is known only from the Seychelles and Aldabra in the Indian Ocean, and it probably has not been seen in the aquarium. Dubbed the "Seychelles clownfish," the species is symbiotic with *Stichodactyla mertensii*.

Amphiprion latezonatus Waite 1900

This species is known only from Lord Howe Island between Australia and New Zealand, and it is a pity that it is not in the hobby, as it is quite distinctive as compared to other clownfish species. It is nearly black in coloration,

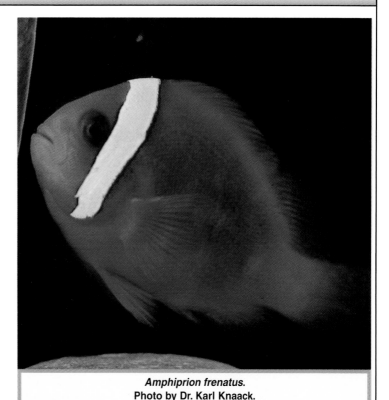

Amphiprion frenatus.
Photo by Dr. Karl Knaack.

Amphiprion frenatus.
Photo by Kok-mans Choo.

Amphiprion fuscocaudatus
Photo by Dr. Karl Knaack.

Amphiprion latezonatus.
Photo by Dr. Herbert R. Axelrod.

and the middle band is particularly broad. Hence the name "broad-band clownfish" has been suggested as a popular name. It is symbiotic with *Heteractis crispa*.

Amphiprion latifasciatus Allen 1972

This species is probably common on the reefs of Madagascar, but it has been little studied and was described only relatively recently. It has unfortunately not been introduced to the aquarium trade, and no information about its life history has been reported.

Amphiprion leucokranos Allen 1973

This is the most recently discovered of the clownfishes. It has been reported only from Madang, New Guinea, and few specimens have been collected. Hopefully, this will be corrected in the future, as this is a quite beautiful little fish. It is known to be associated with *Heteractis crispa*.

Amphiprion mccullochi Whitley 1929

This very unusual little clownfish is nearly black in color, with very distinctive white stripes curving down over the gill cover. It is known only from Lord Howe Island, and it, too, has not yet been introduced to the aquarium trade. As is the case with so many other clownfishes, this species is associated with the anemone *Heteractis crispa*.

Amphiprion melanopus Bleeker 1852

This is a widespread species, ranging from Indonesia to the South Pacific. Like so many other wide-ranging

Amphiprion leucokranos.
Photo by Roger Steene.

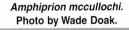
Amphiprion mccullochi.
Photo by Wade Doak.

Amphiprion melanopus.
Photo by Dr. Herbert R. Axelrod.

Amphiprion nigripes.
Photo by Dr. Herbert R. Axelrod.

clownfishes, this one is quite variable in coloration. Some of them are completely orange with only one stripe, and others have the dark body coloration and may have two stripes (although the two stripes are rare in adults). The blackish coloration on the sides of many specimens has given the species the common name of the "red and black clownfish." It usually accepts only *Entacmaea quadricolor* but is also associated with *Heteractis crispa.*

Amphiprion nigripes Regan 1908

This species is known from the Maldive Islands and Sri Lanka. Often confused with the previous species, the Maldive clownfish has been shown to be a valid species by Allen and Mariscal (1971). It is found only with *Heteractis magnifica.*

Amphiprion ocellaris Cuvier 1830

This is the "common clownfish" that has been in the hobby for so many years and often is referred to as *Amphiprion percula.* It is widely distributed, from Thailand to the Philippine Islands, so it is no accident that it was one of the first clownfishes kept (and probably the species that gave the group its name). It is symbiotic with *Stichodactyla gigantea*, *Stichodactyla mertensii*, and *Heteractis magnifica.* The species has been bred many times.

Amphiprion percula (Lacepede 1802)

This species was originally described as *Lutjanus percula.* Its distribution is limited to Northern Queensland, New Guinea, New Britain, New

Amphiprion ocellaris.
Photo by John Manzione.

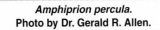
Amphiprion percula.
Photo by Dr. Gerald R. Allen.

Amphiprion percula showing a color variant of a young fish.
Photo by Dr. Fujio Yoshida.

Amphiprion perideraion tending it eggs which are always
laid close to an anemone.
Photo by Dr. Gerald R. Allen,

Ireland, the Solomon Islands, and New Hebrides. Since it is not as widely distributed, it is not surprising that it was not until recent years that it was imported. It takes something of a practiced eye to tell the difference between this "common clownfish" and *Amphiprion ocellaris*, but this one has a slightly higher dorsal and often dramatic black markings along the white stripes. The species has been bred in the aquarium and accepts *Stichodactyla gigantea* and *Heteractis magnifica*; it has been bred without an anemone present.

Amphiprion perideraion Bleeker 1855

This is one of those species known by the popular name of "skunk clownfish" because of the white stripe down its back. It has a wide distribution, from the Philippine Islands to Japan to the Marshall Islands, so it has been collected and been part of the marine aquarium hobby for some time. It may not have the dramatic color of some species, but its tasteful hues, pleasant demeanor, hardiness, and personality have made it an aquarium favorite. The only anemone that it seems to accept is *Heteractis magnifica*.

Amphiprion polymnus (Linneaus 1758)

Although this species was among the first of all clownfish species to be described, it is not common in the hobby. It is variable in coloration, sometimes having an almost black body coloration and other times being quite red. And the stripes are variable, too. Sometimes it

Amphiprion perideraion.
Photo by Dr. Denis Terver

Amphiprion polymnus, at their natural habitat in the Solomon Islands (Guadalcanal).
Photo by Dr. Gerald R. Allen.

Amphiprion rubrocinctus.
Photo by Dr. Gerald R. Allen.

has two stripes and other times three, and the stripes—especially the posterior one—may vary as to position. However, because the second stripe (if there is only one) is usually large and only extends halfway, or slightly more, down the body, the species has the common name of "saddleback clownfish." It is found from China to the Solomon Islands, and it is symbiotic with *Heteractis crispa* and *Stichodactyla haddoni.*

Amphiprion rubrocinctus Richardson 1842

One of the "tomato clownfishes," this species has probably only rarely, if ever, appeared in the tanks of American hobbyists, as its distribution is limited to the Australia region. For that reason, Dr. Gerald R. Allen has suggested the common name of "Australian anemonefish." It is symbiotic with *Entacmaea quadricolor.*

Amphiprion sandaracinos Allen 1972

Found from Western Australia to the Philippine Islands and other western Pacific locales, this may be the original "skunk clownfish," as it does have the white line down the back. All the skunk clownfishes have a reputation of being hardy and easy to keep, and while not as colorful as some of the other clownfishes, they have a charm of their own. This species is symbiotic with *Stoichacatis mertensii.*

Amphiprion sebae Bleeker 1853

Another species that has not been in American aquaria (or, perhaps, even others),

Amphiprion sebae is limited to the southern coast of Arabia, India, Sri Lanka, the Maldive Islands, Andaman Islands, and the Indian Ocean coasts of Sumatra and Java. Since it is unknown as an aquarium fish, there is no common name. In any case, it is usually referred to as "sebae" (see-bee), and that name is as easy to pronounce and remember as any common name. It is symbiotic with *Stichodactyla haddoni.*

Amphiprion tricinctus Schultz and Welander 1953

This species is known only from the Marshall Islands in the Pacific Ocean. It is unlikely to have been collected as an aquarium fish and has no popular name, although Dr. Allen suggests "three-band anemonefish." The species is symbiotic with several anemones, including *Entacmaea quadricolor, Heteractis aurora, Heteractis crispa*, and *Stichodactyla mertensii.*

Premnas biaculeatus (Bloch) 1790

This is the only clownfish species that is not in the genus *Amphiprion*. It is differentiated from the members of that genus by (among other characteristics) a large spine on each side of the head, just below the eye. This species is known among marine hobbyists as the "maroon clownfish," and it is exceedingly popular because it is hardy and easy to keep in pairs. Also, it readily accepts a variety of anemones. In fact, it is recommended that this species not be kept in a mini-reef tank without a host anemone, as it will actually harass the corals, so strong is

Amphiprion sandaracinos.
Photo by Arend v.d. Nieuwenhuizen.

Amphiprion sebae, from Sri Lankan waters.
Photo by Rodney Jonklaas.

Amphiprion tricinctus.
Photo by Dr. Gerald R. Allen.

Premnas biaculeatus from Marau, British Solomon Islands.
Photo by Dr. Herbert R. Axelrod.

An *Amphiprion frenatus* snuggled into an *Entacmaea quadricolor* anemone.
Photo by Arend v.d. Nieuwenhuizen.

its drive to adopt an anemone. It influences the coral polyps to stay open and rubs against them. A little of this goes a long way with the coral polyps, and they soon succumb. This behavior is obviously a "misfiring" of the instinct to adopt an anemone. Other clownfish species are known to do the same thing, but this one is the worst. Surprisingly, the species is found only with *Entacmaea quadricolor* in nature. *P. biaculeatus* is found from India to the Phillipine Islands to the Solomon Islands.

Although most females are the larger of the pair in clownfish species, it is most pronounced in this one, with the female being over twice as large as the male. Allen reports that the species is not maroon at all in its natural habitat but is actually bright red. Experiments in diet and lighting have shown that it is possible to bring out the red coloration in aquarium specimens. In any case, the maroon coloration is spectacular enough, and this species is sure to remain a favorite in the hobby, as it has been bred in captivity.

REMARKS ABOUT IDENTIFICATION

We of course need to know what species we are talking about in order to be sure that we are discussing the same animal. But there are many factors that enter into the picture and blur the identity even of those species with scientific names. There are many color variations, for example, as was mentioned for *Amphiprion clarkii*. Many of these are geographical variations, with some species

Entacmaea quadricolor being cleaned off by almost transparent cleaner shrimp.
Photo by Gunter Spies.

Closeup of the tentacular ends of the anemone *Entacmaea quadricolor.*
Photo by Gunter Spies.

showing variable amounts of red according to where they are found in their range. That is one reason that some "common clownfish" will look so much better than the others. But things get extremely complicated here, because the reason for dull coloration not infrequently has nothing to do with where the fish came from and resuls instead from a poor diet. The only way you could be sure about this, however, is if your specimens were bright in coloration when you bought them and became drab in your tank.

But things get even more weird. (The life of a clownfish devotee may, on occasion, be hectic—but it is never dull!) Dr. Gerald R. Allen reports that at least two species of clownfish will change color if introduced to another anemone. The species mentioned were *Amphiprion clarkii* and *Amphiprion tricinctus*. Allen had noticed that the species varied in coloration (from black to red) depending on the anemone species with which they were associated. So he tried switching fish to different species and discovered it took them about two weeks to make the color change. Presumably, the color change had something to do with camouflage, but it is

Some clownfishes change color when they are introduced to an anemone with which they are not familiar. The fish is *Amphiprion clarkii* in a gold-color anemone with its white bars being pure white. Photo by Dr. Herbert R. Axelrod.

Amphiprion clarkii in a dark colored anemone. Its white bars become bluer and the fish
becomes darker than normal.
Photo by Walt Deas.

really quite baffling. The clownfishes are practically immune from predation while they remain in the anemone, so why would its color influence that of the clownfishes?

It should be mentioned that most anemonefishes live in pairs, with the male being slightly smaller than the female. Often an aggregate of young clownfish of the same species is found with them. It

largest one will do that. The pair somehow prevents the juveniles from maturing. Darwinian pressure has apparently caused the pair to tolerate the juveniles if they are of the same species. The pressures are obvious. If a member of a pair of clownfish were to die, the mate would be left to live out its life alone and unproductive in regard to progeny.

the largest male again would retain its sexual identity, but the male member of the pair would change to a female. (Yes, all of this is very weird from our perspective, but it has been a system that has made the clownfishes a very successful group.)

In any case, you now have a list of species to pick from if you have decided to keep a clownfish display. Among marine aquarists, the most hardy specimens are believed to be the skunk clownfishes, the tomato clowns, and the common clowns, in that order. These are fishes that have been long kept. Some of the rarer species may be just as hardy, but they haven't been in the hobby, or they haven't been in it often enough for there to be a consensus about their hardiness.

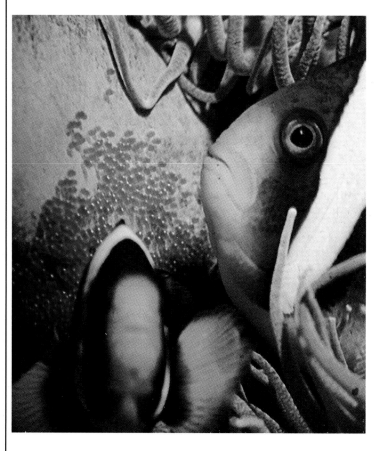

A pair of *Amphiprion clarkii* guarding their red eggs. Photo by Jan Carlen.

might be thought that these were progeny of the adult pair, but such is not the case. Clownfish colonize anemones straight from the planktonic rafts. The largest fish in an anemone become a pair. If they are not a pair, the male has the ability to change to a female, and the

Hence, the juveniles are tolerated *if they are of the same species*. That way if one of the pair dies, the largest of the juveniles takes the place of the missing fish. If it was the male that died or was taken by a predator, the juvenile would retain its male status. If it was the female,

You may decide to keep clownfish as part of a fish-only community tank, as a clownfish and anemone exhibit, or for the purpose of breeding. (If you want to do this, you should keep them as a display first, to "cut your teeth" on them and learn their ways.) I'll describe ways to accomplish each of these.

CARE AND FEEDING OF CLOWNFISH

In the wild, most clownfish species feed upon the same plankton that feeds the anemone. They also supplement their diet with algae scraped from around the base of the anemone, and they will opportunistically take other foods. That is one reason that they feed so well in the aquarium. They thrive on a diet of newly hatched brine shrimp, live brine shrimp, specially formulated frozen foods, and even good quality dry foods.

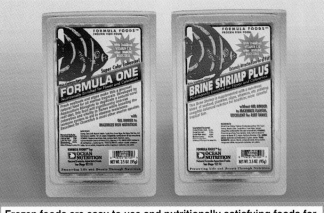

Frozen foods are easy to use and nutritionally satisfying foods for clownfishes and their associated anemones. These products are available at most pet shops specializing in fishes. Photo courtesy of Ocean Nutrition.

The importance of newly hatched brine shrimp can't be overstated. This food is extremely rich in nutrients, as the yolk is still attached, and it very much approximates the food that the fish would eat in the wild. So even the adults can be fed newly hatched brine shrimp as the basic food, with everything else being supplemental. And if you are keeping an anemone, the same food will suffice for it, too. It should be mentioned, though, that there may be essential nutrients that are not present in the newly hatched brine shrimp; hence the need for supplemental marine foods in the form of frozen, dry, and dried foods.

Good nutrition will enhance the color of your fish and keep them in good health. The best way to achieve good nutrition

for your clownfish is to provide them a variety of foods. So it is a good idea to vary your feeding regimen somewhat. For example, you could feed a basic diet of newly hatched brine shrimp. If that is your staple, you could feed it either mornings or evenings. Assuming that you're going to feed twice a day, the other feeding could be either dry food or frozen food, and you could alternate that each day.

Of course, several small feedings a day are ideal. If you have the time for doing so, you could feed small quantities of newly hatched brine shrimp mornings and evenings. You could then feed dry food in the middle of the day and provide a good-quality frozen food for the late evening meal. Occasionally, a fresh leaf of romaine lettuce should be provided. Just place it in the water and remove it an hour or so later. You may have to offer it once or twice before the clownfish

acquire a taste for it, but even that is a rare situation. Although romaine lettuce doesn't grow in the sea, you wouldn't know it from the way marine fish take to it, and it doesn't have to be boiled or frozen either. They seem to prefer it fresh and untreated!

MAKING THE DECISION

Although there are a variety of methods to keep and display clownfish, a couple of main decisions need to be made. First, do you want to keep the clownfish as part of a display in a fish-only tank? If such is the case, prepare lots of hiding places in the tank. Clownfish are seemingly content without an anemone as long as they have a tank in which there are plenty of hideaways. One way to provide this is to utilize artificial corals. There are several advantages to using "synthetic" corals. One is that the corals look like the real thing. If you get real corals, they will just be the skeletons of the corals, and they will be bleached white. (And by buying them, you will be encouraging destruction of coral reefs.) In any case, don't worry about having too many hiding places for the clownfish. They will only use those hideaways part time, and the fact is that just knowing that they have them will make the clownfish all the more bold about appearing out in the open.

If you are already maintaining a mini-reef tank, most clownfish species are good candidates for such a tank, but they are not as good a choice as certain other species. That is because clownfish will sometimes "court" coral polyps, causing them to stay open and possibly eventually killing them from all the attention and rubbing. Not only that, but clownfish have been known to pick at the polyps. It doesn't happen often, but it is a possibility that must be noted.

If you are keeping a mini-reef tank, you may want to think in terms of keeping an anemone, one that the clownfish will accept as a host. Most anemones are sold in aquarium shops under common names, such as "carpet anemone." Most anemones that I have seen sold under that particular name were *Stichodactyla mertensii*; however, the aquarium shop owners usually know which species of clownfish will accept the anemone. The fact is that many clownfishes will accept the same species of anemone. There are hundreds of species of anemones, but only a few are accepted by clownfishes. All of these species are sold as "clownfish anemones." The main drawback of keeping an anemone in a mini-reef tank is that the anemone may wander about, and its sting-

ing tentacles can damage the corals. It is best to place the anemone in the tank first, let it get settled down in its favorite spot, and add the corals later.

If you decide to keep your clownfish in a fish-only tank, undergravel filtration will be adequate. I tend to prefer the reverse-flow undergravel filtration setups, as they keep the particulate matter separate from the biological filter. It has been my experience that this is the best system as long as the mechanical filters are cleaned regularly. The mechanical filters may be either the prefilters for the reverse-flow undergravel filters or the canister filters that pump the water down the lift tubes and underneath the

An anemone *Entacmaea quadricolor* harboring *Amphiprion melanopus*.
Photo taken 30 feet down at Efate, New Hebrides.
Photo by Dr. Gerald R. Allen.

filter plate. If you decide to go with a regular undergravel filter, I have had the best results (and observed the best in other tanks) by using lots of aeration and not using powerheads for pumping the water through the filter bed.

An even better setup involves the use of compound filters. You could have an undergravel filter and an additional biological filter, such as a trickle filter or a fluidized bed filter—or both! The type of tank that I have really fallen in love with is of fairly recent design. Such a tank may be either Plexiglas or regular glass, but it incorporates a large filtration area that is actually part of the tank. The partitioned-off area is composed of several compartments, ready made for compound filtration. There are slits in the back glass that allow the water to flow into the compartment on the left. Then there are slits in the partition glass between the various compartments to allow the water to flow. The reason that the water will flow is that there is a powerhead plumbed right into the back glass in the last partition on the right side, and it returns the water to the tank. The great advantage in this system is that the pump does not have to work hard, as it is not working against gravity. But the biggest advantage of all is that there is no plumbing present that is going to produce leaks on your carpeting, because everything is contained inside the tank. This is no small advantage—especially if you have a spouse who takes a dim view of such things!

If you purchase such a tank, you could have layers of filter floss with bags of activated carbon in between them in the first compartment. Usually there is room enough to include a protein skimmer in the first compartment, too, and the protein skimmer is one of those rare pieces of equipment that actually exceeds expectations. It is not without reason that they have become popular in the marine hobby. In the second compartment you could have a trickle filter setup, and in the third you could have a fluidized bed filter. Usually there are four compartments, and many people simply utilize the fourth for observing the water to see how clear it is by the time it gets there. Also, that compartment can be used to place some of the things that you don't want inside the tank for aesthetic reasons, such as a thermometer, hydrometer, heater, and other instrumentation you may need. This is also a good compartment to place a mark on the glass that shows the water when it is at its optimum salinity. That way you know to add fresh water whenever the water drops the least bit below the mark.

With such a tank, you get the best of all worlds. Although I love simplicity, a tank such as this enables me to have different biological filtration, a protein skimmer, *and* chemical filtration and still be able to easily keep an eye on all that is happening. You can tell that I am enthusiastic about such a setup. The beauty of it is that the only thing that is ever going to wear out is the powerhead, and that is easily replaced.

The other main decision you need to make about keeping clownfish is whether you want to keep an anemone with them or not. The fact is that the anemone will be much more demanding in regard to water quality requirements than your fish. And it will need special lighting in order to keep the symbiotic zooxanthellae in its tissues. Fortunately, such lighting has become quite proficient and even affordable. The time once was that a tank that provided such lighting had about six tubes in the hood, and even they were supplemented by hanging halide lights. Now many specially made hoods have just two fluorescent tubes; one is normal daylight lighting, and the other provides the actinic lighting. In addition, fans are built into the hood to keep a flow of cooling air going through the hood. Needless to say, I am enthusiastic about these new aquarium covers, too.

In addition to high intensity and special spectrum lighting, you will need to provide your anemone with super clean water. You should probably have a protein skimmer and a good biological filter, such as a trickle filter or a fluidized bed filter. The latter is fairly new to the aquarium scene, but the fact that it has become so popular so fast is an indication that it does a pretty good job of providing biological filtration.

PREPARING THE CLOWNFISH AND ANEMONE DISPLAY

If you have decided to keep your clownfish with an anemone, you are to be congratulated, for it is one of the most beautiful and interesting of all possible exhibits. But having the anemone does complicate things somewhat. Remember, you will need special lighting to successfully

Amphiprion perideraion make an interesting contrast to this green anemone which has not been identified.
Photo by Walter Deas.

keep the anemone, and your water quality must be near perfect. Also keep in mind that if anemones do not prosper and start to die, they can foul the tank beyond belief. That is extra incentive to keep a happy and healthy anemone! Because of the sensitivity of anemones, it is best to get them situated in the tank first, before you put your clownfish in the tank. If your anemone is prospering, you know that the clownfish

will be okay—but, of course, you will want to quarantine them before placing them in the tank with the anemone.

If you are going to be keeping an anemone, you need to make some decisions about gravel in the tank. What follows is information for something to keep in mind as possible experimentation after you have cut your teeth keeping clownfish and anemones in the tried and true manner.

An approach that seems to work well is to make use of living sand. Living sand is coral gravel that has been taken from the coral reef and transported in such a way that all the organisms that live in the sand are kept in good shape. These organisms include beneficial bacteria as well as crustaceans, sea cucumbers, and brittle starfish, all of which help keep the gravel stirred up and cleaned up. You don't need to place all living sand in the bottom of your tank. You can place (preferably) aragonite sand that is not living in the bottom of the tank, then place in enough living sand for the bacteria and other organisms to colonize the non-living sand that has been placed there.

Living sand is part of the "natural reef" approach, which is becoming quite popular, and reports on it seem too good to be true. Yet it is done often enough to know that it can work. However, the people who have been successful at it have been extremely experienced hobbyists. It must still be considered in the experimental stage, so I am just throwing this idea out for future consideration.

The ideal unit for the clownfish aquarium is the protein skimmer combined with a biological filter.
Photo courtesy of the Cyclone Bak-Pak.

The anemone *Stichodactyla haddoni*.
Photo by U. Erich Friese.

Amphiprion nigripes from Sri Lanka.
Photo by Rodney Jonklaas.

In the meantime, let's plan on utilizing at least a protein skimmer and trickle filtration, along with chemical filtration from the bags of activated carbon that will be placed in the sump of your trickle filter. (There are also protein skimmers designed to go in the sump and be out of sight and not require separate plumbing for them. These protein skimmers work well enough that they deserve consideration on these merits alone.) For gravel, you could have a sprinkling of living sand. In the natural system, competing biological filters and protein skimmers are shunned, as all the metabolic waste is supposed to be consumed by the living sand, live rocks, and other organisms. In theory, this is great, but as I mentioned, the natural approach must still be considered experimental. If you have just a sprinkling of living sand, it is not going to be starved by your trickle filter and protein skimmer. Neither is it likely to add significant pollution to the tank.

Another option would be to place a reverse-flow undergravel filter in the tank. (But don't do that with living sand!) That way you could have a normal bed of gravel, and the water quality should still stay high as long as the sand is regularly vacuumed and the mechanical filter part of the reverse-flow system is regularly cleaned.

An anemone *Heteractis crispa* hosting *Amphiprion clarkii*.
Photo by Dr. C.W.Emmens.

Whatever system you use, you will need to allow the tank to cycle. That means that you need to allow the biological filters to reach their full working capacity. It takes time for the beneficial bacteria to colonize the medium in your trickle filter and the gravel in your tank. As mentioned earlier, many people use inexpensive damsels to provide the metabolic byproducts that will serve as "food" for the biological filter. In the normal course of events, the ammonia and nitrite levels will climb and then begin to drop dramatically. This process takes from two to three weeks.

When the measurements of your water parameters are all optimum, you can place the anemone in the tank. Have your tank laid out in the manner you plan to keep it, as the anemone may wander a bit before settling down. Fortunately, it will normally stay in one place once it finds a favored spot. You should have a timer on your lights so that the anemone gets sufficient lighting. A twelve-hour cycle of having the lights on and then off is preferred, as that is the approximate length of the tropical day. Naturally, you may wish to have the twelve hours include the time of day (the evening, usually) when you may wish to view the tank. This can easily be accommodated, as the best place in the house is a darkened area, which allows you to have complete control over the lighting. In any case, a little extra natural daylight is not going to be harmful to the anemone.

I would advise you to keep the anemone by itself (by that I mean without clownfish, as the damsels used for priming the biological filters may still be kept in the tank) for at least two weeks. Yes, there is

A *Condylactis* anemone eating a large *Chaetodon triangulum*.

a need for patience in this hobby! The reason for this is to make sure that the anemone is prospering.

When you finally do place your clownfish in the tank, there is the danger that they will immediately dive into the anemone without acquiring the molecules from the anemone to keep from being stung, and they may actually be stung to death. The reason for this danger is that the clownfish will be frightened (as all fish are) by being transported in the plastic bag, and upon release, their instinct is to dive for safety. They will instinctively seek the anemone. This is a different situation from juvenile clownfish descending from the plankton raft. In the latter situation, the clownfish seemingly flirts with the anemone, approaching with caution and allowing itself to be touched only tentatively by the tentacles of the anemone. It doesn't take long, but it does take some time for the clownfish to build up the layer of cells in its specialized coat of protective slime that, in effect, turns off the stinging cells of the anemone. In a stress situation, however, the clownfish "forgets" about that process and simply dives for cover in the anemone, and such a dive can very well be fatal for it.

To prevent such a thing

from happening, the proper way to introduce a clownfish to a new anemone is to place plastic grating close to the anemone. This keeps the clownfish out of the anemone but allows it (or them) to "flirt"

Clownfish aquariums, as well as reef aquariums in general, benefit from additional calcium that stabilizes alkalinity without affecting pH, releases carbon dioxide to enhance photosynthesis in the zooxanthellae of the anemones and corals, and is biologically safe.
Photo courtesy of Tropic Marin USA, Inc

with one or two protruding tentacles. After only half an hour or an hour, the grid may be removed, and things should be fine from that point on.

LIVE ROCK

If you are going to keep an anemone with your tank, you may wish to make use of live rock. The question is: What is live rock? Live rock is rock

that has been colonized by desirable bacteria from the ocean and by coralline algae and other desirable sea life. These rocks used to be taken from the ocean, but many such rocks are now cultivated in aquaculture. The rocks may either be placed in the ocean to allow them time to be colonized or they may be cultivated in various large aquaria. If you place suitable rocks in a large tank already full of healthy live rocks, your inert rocks will soon be colonized, too, and that is the principle many aquaculturists operate under.

The reason live rocks are cultivated somewhat artificially now is because of government restrictions. But it is good for the hobby in the long run, for commercial dealers are learning to provide us with better live rocks. That is, they can provide us live rocks without the pests, such as bristle worms and mantis shrimps, that can wreak havoc in a tank.

Examples of Clownfish Displays

The following are displays that I have either had good success with myself or observed with others. They are all of different size with a different match of "characters." They are provided to give inexperienced hobbyists some possible ideas for display.

THE 5-GALLON TANK

For lighting, there are compact fluorescent bulbs and fixtures that are manufactured for such tiny tanks. The very fact that they are manufactured demonstrates that a small tank is not such an unworkable thing. Get one blue actinic bulb and one daylight bulb. It is good to have these on a timer and set it for 12 hours of daily light. (You may want to set the timer, however, so that the light is going to be on in the evening, too, so that you can view the tank at that time.) You will need about eight pounds of live rock. That means that you can get six pounds of base rock. You will need only two pounds of true live rock, so you will be able to afford the very best.

For equipment, get an outside filter that hangs on the back and incorporates some trickle (or fluidized bed) filtration and an inside protein skimmer that fits in the filter. The nice thing about a small tank like this is that the filter systems easily handle such a small volume of water. Your filter may be nearly as large as your tank, and that effectively doubles the water that you are actually keeping the fish in. You can keep your heater and your thermometer in the outside filter. Check both of them daily as you check on your protein skimmer and empty its collection cup (and check to make sure that the bubbles are flowing correctly).

At this time, you can add your invertebrates. You may have some coralline algae on your live rock and some other growth. I suggest adding a small orange sponge down in a darker portion of the tank and some green star polyps. The star polyps will begin to colonize your live rock. (They will colonize the inert rock, too, which then will eventually qualify as full-fledged live rock.) To control unwanted growth of micro-algae, introduce a turbo snail.

After a couple of months with everything going well, add a small (about four to five inches across) clownfish anemone to the tank. After the anemone has settled down and seems to be prospering, you can think about adding fish.

It will probably take six weeks for the rock to accli-

Amphiprion nigripes in their natural habitat in the Maldive Islands.
Photo by Dr. Herbert R. Axelrod.

mate to your satisfaction.
Keep track of your water
qualities with your test kits. If
everything is okay, add a
couple of clownfish. Which
species you add is not too
important as long as the fish
are small. Common
Amphiprion percula or *A.
ocellaris* look great. Maroon
clownfish, *Premmas
biaculeatus*, are good because
they readily accept nearly any
anemone, and they are easy
to purchase as pairs. The only
problem with that species in
such a small setup is that the
female can attain a size that

Opposite page: *Amphiprion clarkii* takes on a very light color when
associated with this tan anemone.
Photo by Pierre Laboute.

Below: *Amphiprion akindynos* amongst living rocks.
Photo by Dr. Leon P. Zann.

might be a little large. (In clownfishes the female is the larger fish.)

Maintenance will consist of checking the equipment daily, adding dechlorinated fresh water daily, and adding trace elements weekly. Of course, check your protein skimmer daily and make sure the other components of the filter are doing their job and are operating properly. (This quick check will be part of the maintenance schedule of any tank regardless of size.) Mix up about five gallons of synthetic salt mix and keep it handy to draw from for regular partial water changes. A good schedule would be to siphon about half a gallon weekly, at the same time siphoning out detritus, and replace it with new sea water. Also, clean the front glass every time you do this. Even for a tiny tank such as this, it is a good idea to keep a log.

You'll be surprised how much fun you will have and what you will learn from a small inexpensive setup such as this, and it will be the talk of all the people who are privileged to see it. The point is that such a small clownfish tank can be kept and still look great. You just have to be aware that changes will have a much quicker effect on it, so you must be nothing short of completely diligent about replacing evaporated water and keeping an eye on all the parameters of the tank, from temperature to pH to nitrate.

THE 30-GALLON TANK

Actually, this tank could be either a 30-gallon or 40-gallon tank, but let us this time think in terms of a tank with the built-in partitions for the placement of compound filters. In the receiving compartment (of the filter), place layers of alternating filter floss and bags of activated carbon. There is usually room for an inside protein skimmer, so place a good quality protein skimmer in this compartment also. Have the next compartment composed of layers of filtration media for trickle filtration. If we really want to have wet-dry trickle filtration, we will purchase and place a trickle plate to be placed just below the slits in this compartment. This can be a little tricky, because if the trickle filter gets clogged or is "rigged" for "slow," the plate can overflow. But most designs provide for any overflow into the compartment. The fact is that we can get good biological filtration with just the filtration spheres without the trickle plate because of the fact that there is such good gas exchange at the surface, as it is completely uncovered. To be really plush, we'll place a fluidized bed filter in the last compartment, thus providing a double whammy at breaking down ammonia and nitrites into nitrates. Again, we have the room, so if we want to be really fancy we can place a denitrator in this compartment, too, and it will help break down the nitrates. This is especially important, as anemones are particularly sensitive to nitrate levels. Place the heater and the thermometer in this section, too—don't forget to keep a good eye on both of these items just because they are back in the filter!

For our tank, we will place about an inch of aragonite sand and seed it with about eight pounds of good quality living sand. We'll leave the tank dark for three or four weeks, but we will place small amounts of dry and frozen food into the tank. The purpose of the food is to feed the tiny organisms that live in the sand and to promote the growth of beneficial bacteria. Keeping the tank dark is to prevent the growth of competing algae. That is, the algae would compete with the growing colonies of beneficial bacteria, and growths of hair algae can become a problem at this stage anyway.

In this tank, we plan to keep at least one clownfish anemone, so we want to have high-intensity and full-spectrum lighting. Fortunately, the type of aquarium we are talking about here offers the option of a hood containing two tubes that provide the proper lighting for just such a tank. Such hoods also have built-in fans to disperse the heat from the lights, and this action also helps promote gas exchange over the surface of the water.

The powerhead that returns water to the tank has a device on it to direct the flow of the water. Naturally, the flow is going to be directed toward the left (as you look at the

The clownfish *Amphiprion melanopus* being cleaned by a *Labroides dimidiatus*. Photo by Allan Power.

front of the tank), but the adjustable nozzle allows you to make fine adjustments within a limited range. The idea is that most anemones and corals prosper best when placed in a current, as that is the way they live in nature. The current helps keep the anemone clean of debris and facilitates gas exchange through its tissues, and there may be other unknown benefits, too, as a current of some kind really helps an anemone prosper. This single-direction type of current is called a "laminar flow." Once you have put your anemone into the tank (after having placed in any rockwork and artificial corals that you want to use for decoration and hiding places for the fishes) and it has settled into place, you adjust the flow of water to come just above the anemone. If the anemone is some distance from the outlet, the flow can be adjusted directly toward it.

After the anemone has been in the tank and has prospered for at least two weeks (and preferably a month), you can place the clownfish in the tank. Let's pretend that you have decided to go with one of the common clownfishes. In that case, your best bet is to purchase about six or seven juveniles. You are then almost 100% sure to end up with a pair (assuming that the tank from which you purchased the juveniles contained roughly a 50/50 mix of males and females to begin with), but the remaining juveniles will be subordinate to the pair and will remain in appearance and actions as juveniles.

After the clownfish have settled down, you can think

about placing other fishes in the tank. A yellow tang will provide contrast, and so will about six small blue damsels. The point is you can have nearly any fish you want (that won't hurt the anemone), but be aware that the clownfish will keep them a respectful distance from the anemone. I have many times been much amused when I have wanted to adjust an anemone's position to have the clownfish ferociously attack my hand. Of course, these little fish can't do any real damage, but their fierce bites do hurt. And if they catch you by surprise, they can most defintely make you jump. (It is much more amusing to watch this happen to other hobbyists!)

THE 200-GALLON TANK

If you really want to do it up big, you can have a large tank that contains more than one clownfish anemone. You should have good biological filtration and the proper lighting. The fact is that you can order an aquarium made with the built-in filter partition on the back, but don't let my preferences influence you! For a substrate this time, let's just use coral sand and utilize a reverse-flow undergravel filter. (Of course, if you have already tried living sand in smaller tanks and have been successful with that medium, ignore what I just said!)

In any case, with the proper lighting and filtration you can place a clownfish anemone at each end of the tank, and you

Zebra moray eels get along well with clownfishes.
Photo by Walter J. Brown.

can have a different species of clownfish in each anemone. A good contrast is a pair or group of common clownfish (either *Amphiprion ocellaris* or *A. percula* or even *Premnas biaculeatus*, although the maroon clownfish is not usually thought of as one of the common ones) in one anemone and a pair of tomato clownfish in an anemone at the other end of the tank.

As the other fish in the tank you could have three blue tangs, one bird wrasse, six blue damsels, one Koran angel, and a zebra moray eel. Such a tank would truly be dramatic, and there would be plenty of room for the anemones to grow. Fortunately, it takes many years for them to grow, as they can reach a size of nearly three feet across, but this is rare to the vanishing point in the home aquarium.

Gomphosus varius, the bird nose wrasse, seems to get along well with anemones and clownfishes.
Photo by Walter J. Brown.

MAINTENANCE

Regardless of the size tank you decide upon or the filtration units utilized, there is some basic maintenance that should be done. It should become a practice to simply look the tank over each day. Make a note to yourself to check each specimen in the tank and observe how it is doing. A really good practice is to keep a log book. This will get you in the habit of looking your tank over with a practiced eye. List anything in the log book that is out of the ordinary. Also, make a note of any time that you make any adjustments of any kind in the aquarium. Include in your log the temperature and the results of any tests you do with various kits for measuring the pH, ammonia level, nitrite level, and the nitrate level.

Make partial water changes on a regular basis. A schedule that has worked well for me has been to change part of the water every other week and add trace elements the week that I don't change the water. Naturally, all of this goes into the log book. If you maintain good aquarium practices, you will some day find that your clownfish have spawned, and that will undoubtedly inspire you to try to raise the babies. That is much easier said than done, but it is now possible, and we will tell you how it can be done as the next order of business.

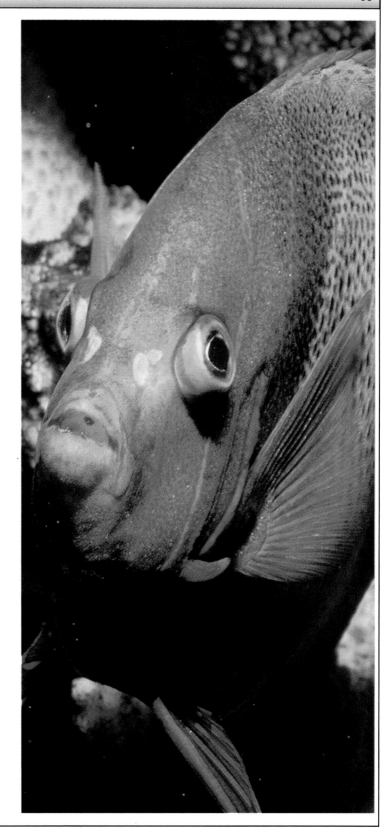

Pomacanthus semicirculatus gets along well with clownfishes.
Photo by MP & C Piednoir.

BREEDING CLOWNFISH

My eyes nearly popped out the first time I saw tank-bred clownfish juveniles and saw the various rearing tanks, as well as the parents guarding young. That was nearly twenty years ago, but I still remember the way a pair of tomato clowns would attack a hand placed in the tank near the eggs just as a pair of cichlids would do! This was at the original Sea World in San Diego, and I and a fellow science teacher were being shown a setup that was generally not shared even with other enthusiastic tropical fish hobbyists. The reason was that marine fish breeding was in its infancy and also was very competitive, so we were asked to not divulge the details of what we had seen. But it certainly got me to thinking about trying my hand at it—even if I didn't have the resources of Sea World.

The reason for my complete surprise was that I had been one of those pessimists who believed that it would be impossible to breed marine fishes. The reason for that was that I was aware that nearly all coral reef fishes, the really desirable ones for the home aquarium, had a pelagic larval form. Almost no known marine fish species gives parental care to their young after they have hatched. And only a few of them even take care of the eggs. One exception is the seahorse, which is famous not only for its appearance but also for the fact that the male incubates the eggs in his pouch. The well developed

young are expelled from the pouch by muscular contractions on the part of the male. Although seahorses lack the pelagic larval form, the parents give no further care to them once they are expelled from the pouch.

**A tank filled with tank-raised juvenile *Amphiprion percula*.
Photo by Don Malpass, Jr.**

The only marine species known to give post-hatch care to their young are the cottontails, *Acanthochromis polyacanthus*, and the Banggai cardinalfish, *Pterapogon kauderni*, which both lack the pelagic larval form. The cottontail is a damsel, closely related to cichlids, and it guards the young for a prolonged period after they hatch, much like cichlids. The Banggai cardinalfish incubates the eggs and young in the mouth

of the male, and the male guards the young and will take them into its mouth for protection for several days after hatching.

Other species that lack the pelagic larval form are the sharks, which often give birth to live young or lay very large eggs. In either case, the young are quite well developed when they arrive into the world, and they receive no parental care. Other examples of non-pelagic young would include the surfperches, which also bear well developed young alive.

Most of the marine fishes don't provide parental care even for the eggs (and, of course, that statement could be made for freshwater fishes, too, but it may be even more applicable for marines).

A breeding pair of *Amphiprion percula*. The male is guarding the blue eggs. The female is
in the anemone while a juvenile approaches from the rear.
Photo by Don Malpass, Jr.

However, the jawfishes brood the eggs in their mouths, as do the cardinalfishes, and the males of many blennies, wrasses, and damsels are known to guard the eggs, sometimes even providing a nest for them. But once the eggs hatch out, they are dispersed to the plankton rafts by the currents. Trigger-fishes are unusual in that the guarding of the eggs falls primarily to the female, who keeps the eggs clean and aerated by blowing water over them—and she fiercely at-tacks any intruders, including divers. However, the eggs hatch quickly, and the young are believed to have a pro-tracted larval form. Although triggerfishes have spawned in the aquarium, there has been no known success at raising the young.

Because it would be so difficult to reproduce the exact conditions of the plank-tonic environment, I, along with most marine biologists, was very pessimistic about marine species ever being raised for the aquarium. How wrong I was! Many marine species have now been raised, and the easiest of all are probably the clownfishes.

Of course, "easy" here is a relative term. It is still very difficult as compared to almost all freshwater species. First, you must have compat-ible adults in good condition. You must have a good envi-ronment with good quality water, and you must be able to transfer the young once they hatch out into a rearing tank with water of the same temperature and chemistry. You need a tank of at least twenty gallons in capacity for the spawning pair, and you

need a tank of about forty gallons in capacity for rearing the young. And then you will need a first food for the clownfish young, as they will be too small for newly hatched brine shrimp. The marine rotifer *Brachionus plicatilis* has proved suitable for use until the young are large enough to take newly hatched brine shrimp, a time of about two weeks. Even finely crushed dry food has been used, but because the number of fry raised is greatly decreased thereby, that is something to keep in mind only if for some reason your rotifer culture goes bad.

The best way to get a pair of parental clownfish is to raise them yourself from juveniles. You are way ahead in the game if you can get tank-raised juveniles. They may be more expensive, but they will be free of parasites; additionally, they will have been born of parents that reproduced themselves in an aquarium, and there just may be a hereditary trait in regard to propensity to spawn under artificial conditions. (It is worth noting, for example, that tropical fish hobbyists have great difficulty with spawning certain wild-caught freshwater fishes, whereas the tank-raised individuals of the same species are considerably easier. The fish from the wild still retain the reluctance to spawn in the aquarium.) If you get tank-raised clownfish, it may take as long as eigh-teen months for them to reach sexual maturity.

If you don't have the pa-tience to raise the young, it is certainly possible to get a pair of wild clownfish. The maroon clownfish, *Premnas*

biaculeatus, is particularly easy to buy as a pair, as they pair up easily, even in a dealer's tank, and they are often collected as pairs. The female is quite large in this species, although the male remains small, so if small size is a consideration, you may wish to stick with one of the common clownfishes. Even here, the male is smaller than the female, however. In any case, it is possible to buy mated clownfishes of many different species.

While you don't have to have an anemone to spawn the common clownfish, you are again ahead of the game if you do keep an anemone with the clownfish. This makes sense. The fish feel more at ease, as they spend their entire lives in the anemone or near it in the wild. In the wild, the fish spawn on a bit of the substrate so near the anemone that the tentacles may cover the spawning site. In fact, the clownfish have been known to bite the ten-tacles to "train" them to withdraw from that particular site. In any case, the point is that it would be to your advantage to keep an anemone with a potential spawning pair. It will bring the day closer that they do spawn, and it apparently improves the mental well-being of the clownfish.

Whether or not you keep an anemone with your pair, you should not keep any other fish in the tank with them. Those who have spawned clownfish have noted that spawning is much likelier to occur when the animals are given privacy in the aquarium. One reason for that fact is probably that it is

A pair of *Amphiprion frenatus* guarding their eggs.
Photo by Dr. Herbert R. Axelrod.

difficult for the pair to achieve the minimum spacing between them and other fishes under aquarium conditions. Another factor is that the anemones in the wild that are populated by clownfish are usually much larger than those that we have in our aquaria, so that automatically gives them a minimum separation from other fishes.

Since you are going to be keeping an anemone, you will have to have the high-intensity and actinic lighting, along with super-good water. The timer for the lights may be set for at least a 12-hour day, and some hobbyists have discovered that a 14-hour lighting time is preferable for eliciting spawning. The water quality should not be difficult to maintain, as you are only keeping an anemone and two clownfish in a 20-gallon (minimum!) tank. A protein skimmer and trickle filter (or fluidized bed filter) will help out on that end. In fact, you may have one of those compartmentalized tanks of which I am so fond and thus be able to indulge easily in compound filters. That is not a bad idea, as water quality is very important here. One way to ensure high-quality water is to make frequent partial water changes. Instead of the once-every-two-weeks changes, it would be a good idea to make them daily and add the trace elements once every two weeks.

In order to condition the pair, you will want to make sure that they receive adequate nutrition. That can be done with a morning feeding of baby brine shrimp and a feeding of good-quality frozen food that has some plant material in it. Make the last feeding either a small offering of adult brine shrimp or of a top-quality dry food. Three

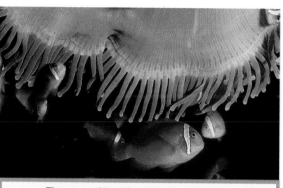

The ease with which *Amphiprion* species interbreed has caused some scientists to be concerned with the taxonomy of the group. These are hybrid youngsters between *Amphiprion ephippium* and *Amphiprion frenatus*. Photo by MP & C Piednoir.

feedings a day would be great and four would be even better. The thing to remember here is that in the wild, the clownfishes are basically omnivores and browse throughout the day on the plankton that is brought to them by the same currents that feed the anemone. But they also browse on items they find on the substrate, including bits of algae. So duplicating that environment by frequent feedings is a good idea as long as the supply of food is extremely light. Remember that we have to keep that water quality at near perfection. Don't worry about feeding the anemone, as it will get some of the newly hatched brine shrimp and some of the other foods, too. I have seen

hobbyists who insisted on feeding the anemone with a turkey baster, and the main thing that did was cause the clownfish to attack the meat baster and hand, as they have a strong instinct to defend their territory, which in their case happens to be the anemone.

Remember that you are also maintaining a 40-gallon aquarium for the larvae to be raised in. You can't be blamed if you don't want to keep that going before the pair have spawned. There are two approaches here. You don't want filtration in the rearing tank for the first few weeks, because the fry will be so tiny. So you can get by with air stones. The water won't degrade while you are keeping the tank empty, but you may lose some trace minerals, so you will have to maintain it in that way (and by adding fresh water to replace what has evaporated). Of course, you will be maintaining the temperature of the water in the rearing tank the same as in the spawning tank, and that should be eighty degrees Fahrenheit. The other method is to keep the rearing tank dry. When the pair spawns, you will have seven to eight days before the eggs hatch, so you can take a little water out of the spawning tank each day and fill it in that way. That way the spawning tank gets lots of partial water changes, and the rearing tank will have water identical to that of the spawning tank.

A sign that the fish are getting ready to spawn is when the fish begin cleaning a rock near the anemone. There will be lots of chasing back and forth and maybe even a little bobbing up and down, which ethologists call "signal jumps." This activity can last for days or weeks and has driven many a hobbyist nearly crazy while waiting in anticipation of the spawning. To compensate for those times, some clownfishes catch their owners by surprise, having spawned with little courtship or cleaning behavior.

Finally, the female will lay the eggs, and the male will follow along, fertilizing them. The male will be the primary guardian of the eggs, although the female will help to some degree. The male will spend nearly all of his daytime activity guarding the eggs and fanning them, occasionally mouthing them. All of this is believed to keep the eggs clean and to keep them well aerated. At night, both fish

Above:
Amphiprion clarkii **with eggs.**
Photo by Jan Carlen.
Below:
Amphiprion ocellaris
guarding eggs.
Photo by Jens
Meulengracht-Madsen.

appear to sleep in the anemone, and it is believed that that is the case in the wild also. Darkness tends to protect the eggs from would-be predators. Some ichthyologists believe that the male tends the eggs at night but is easily "spooked" back into the anemone. In any case, the nighttime care of the eggs is minimal when compared to the daytime activity.

The eggs will be orange in color, but they will change slightly as they begin to develop over the next eight days. On the day of hatching, the eggs develop a violet sheen over a brown coloration that they have attained.

The eggs always hatch at night, and there is good reason for this. The reef is full of plankton feeders, and the young become part of the plankton as soon as they

Amphiprion clarkii, a breeding pair, with the male being closer to the eggs than the female. This 1979 photo indicates how long clownfishes have been spawned in the aquarium.
Photo by Jan Carlen.

hatch. Even their parents will eat them. After all, they are plankton feeders, too, and they seemingly do not recognize the young when they hatch. Whatever the case, there is good reason for the eggs to hatch in the cover of darkness.

If we anticipate the night that the eggs will hatch, we can darken the tank earlier, and that way we won't have to stay up until midnight waiting for the eggs to hatch so that we can transfer them to the rearing tank. Since the larvae are attracted to light, a flashlight can be utilized to get the young in a concentrated area,

and they can be siphoned out into a container to transport to the rearing tank. Obviously, you wouldn't try to net them out, as even cichlid fry are too fragile for that, and the clownfish larvae will be much smaller than any cichlid fry (although they are large when compared to most other marine larvae).

An important point here is that you will need overhead illumination in the rearing tank. One of the problems is to keep the fry out in the open water. In fact, some hobbyists darken the sides of the tank so that they are not attracted at all to any light reflecting

from the sides. As planktonic larvae, the fry seemingly don't recognize barriers and will bang into the sides of the tank. (As open ocean inhabitants, the larvae had no need to evolve dealing with such things as barriers, and we see the same behavior in many pelagic fishes that are kept in aquaria—even in very large tanks.)

You won't have to begin feeding the fry until about the eighth day. But you'll have to make preparations before that to make sure that you have a good culture of rotifers. Here are some methods for cultivating them.

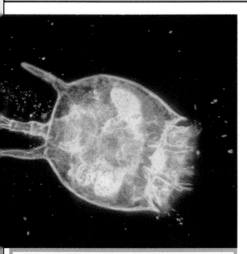

Brachionus are the best food for clownfish fry. They are fairly easy to culture in your own home.

First, since the marine rotifer *Brachionus plicatilis* is such a good first food for marine fishes, commercial enterprises now sell it as a frozen food. The frozen rotifers can be used as food for clownfish fry, but the larvae will eat them only if they are suspended in the water. They ignore them once they drop to the bottom. That means you need active aeration (which you should have anyway in the rearing tank), and you need to siphon off the bottom a few minutes after feeding. Also, you will need to feed several times a day. Obviously, you are not going to get as good a growth rate as you will by culturing and feeding live rotifers. However, you may be tempted to freeze quantities of the rotifers yourself when you raise them, as the cultures rapidly increase in size and then die from their own waste products.

One way to feed rotifers is to feed them algae. Yes, this means that you must culture food to feed the food! Check with your aquarium dealer about obtaining cultures of *B. plicatilis;* these cultures are commonly available now, since they have been so successful in the rearing of marine fishes. Also available are algae spores for good marine species of algae for cultivating rotifers. These include *Dunaliella, Isochrysis,* and *Monochrysis.* The algae can be cultured in gallon jugs or 5-gallon tanks, but they must be kept under heavy illumination and provided

Amphiprion melanopus. Clownfishes are notorious eaters of their own fry once they become free-swimming. Photo by Ken Lucas.

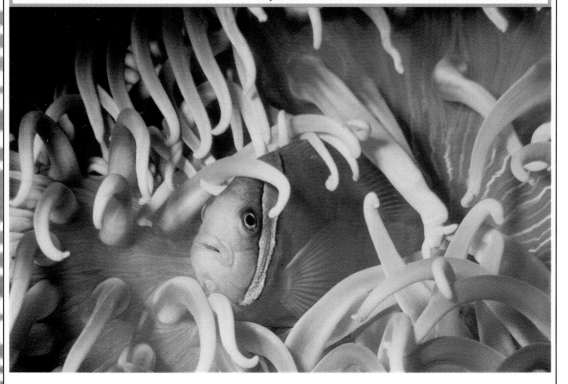

An *Amphiprion chrysopterus* egg at 36 hours after being fertilized.
Photos in this series by P. Frankbonner.

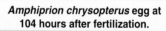

Amphiprion chrysopterus egg at
104 hours after fertilization.

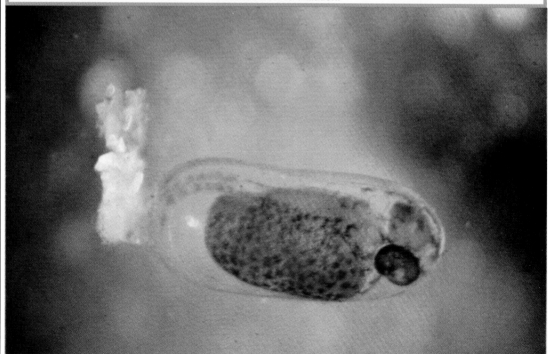

Amphiprion chrysopterus egg at
126 hours after fertilization.

Amphiprion chrysopterus egg at
148 hours after fertilization.

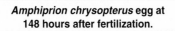

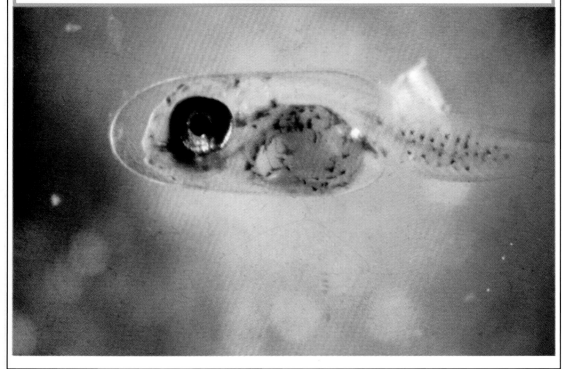

Amphiprion ocellaris tending eggs.
Photo by Jens Meulengracht-Madsen

with "food." Some of the all-element plant foods will work for this purpose, but only tiny amounts should be introduced into the water. The algae can be fed to the rotifers by means of a turkey baster.

Another way to cultivate rotifers is by feeding them cultures of yeast. Even dry yeast will work, but it is not as desirable as either cultured yeast (in warm water under sunlight to discourage bacteria growth) or algae.

The rotifers need a tank themselves of about forty gallons. Bigger is better. A child's plastic wading pool can be utilized in the summer, when the water will be kept warm enough. The sun will help keep the algae going, but you will still need to keep up daily feedings, as the rotifers are ravenous and quickly multiplying. Obviously, you must be certain that nothing can contaminate the pool, such as insecticides or fertilizers.

To feed the clownfish larvae, use a net for harvesting the rotifers. You don't want to take a chance on transferring the water with the animals, so use a very finely meshed net for transfer to the clownfish rearing tank.

When the clownfish have reached a size of a quarter of an inch, they can be transferred to normal tanks with proper filtration for the best growth. By this time, they will be easily eating newly hatched brine shrimp, and you can begin to start feeding other foods, such as tiny bits of dry food and frozen foods to start getting the fry acclimated to a normal diet.

Once your fry have attained juvenile status, you will have no difficulty selling them to an abundance of dealers. But more important than the money is the pride you will have in joining the ranks of the elite: those who have successfully spawned and raised marine fish. The immense satisfaction in that may surprise you. And it will most certainly repay you for all of your hard work.